Nature School

LET'S LOOK FOR SNAILS!

I0794152

By Seth Lynch

Gareth Stevens
PUBLISHING

Please visit our website, www.garethstevens.com. For a free color catalog of all our high-quality books, call toll free 1-800-542-2595 or fax 1-877-542-2596.

Library of Congress Cataloging-in-Publication Data

Names: Lynch, Seth, author.
Title: Let's look for snails! / Seth Lynch.
Description: Buffalo, New York : Gareth Stevens Publishing, [2024] | Series: Nature school | Includes index. | Audience: Grades K-1
Identifiers: LCCN 2022045125 (print) | LCCN 2022045126 (ebook) | ISBN 9781538286296 (library binding) | ISBN 9781538286289 (paperback) | ISBN 9781538286302 (ebook)
Subjects: LCSH: Snails–Juvenile literature.
Classification: LCC QL430.4 .L96 2024 (print) | LCC QL430.4 (ebook) | DDC 594/.3–dc23/eng/20221012
LC record available at https://lccn.loc.gov/2022045125
LC ebook record available at https://lccn.loc.gov/2022045126

First Edition

Published in 2024 by
Gareth Stevens Publishing
2544 Clinton St,
Buffalo, NY 14224

Editor: Kristen Nelson
Designer: Claire Wrazin

Photo credits: Cover Joyce LCY/Shutterstock.com; p. 1 nrey/Shutterstock.com; p. 5 whitejellybeans/Shutterstock.com; p. 7 Derren/Shutterstock.com; pp. 9, 24 (shell) sasha pepper/Shutterstock.com; pp. 11 (main), 24 (foot) Zuzha/Shutterstock.com; p. 11 (arrow) Lyudmyla Ishchenko/Shutterstock.com; p. 13 Oleksandr Smushko/Shutterstock.com; p. 15 Enadin/Shutterstock.com; p. 17 ESOlex/Shutterstock.com; pp. 19, 24 (feelers) Alexander Mak/Shutterstock.com; p. 21 Tsesar Anna/Shutterstock.com; p. 23 Aliaksei Zverkou/Shutterstock.com

Printed in the United States of America

CPSIA compliance information: Batch #CS24GS: For further information contact Gareth Stevens, New York, New York at 1-800-542-2595.

Contents

I see a trail of slime.
Let's look for snails!

They have a soft body.
They have a hard shell.

They hide in the shell.
It is their home.

They crawl.
They use one foot.

They like warm,
wet weather.

They dry out in the sun.
They come out at night.

We see a snail.
It is in the garden.

It has two eye stalks.
It has two feelers.

It makes slime.
This keeps it safe.

It eats the
cucumber plant!

Words to Know

feelers

foot

shell

Index